Tommy's Embarrassing Day

By Cleo Hall

One hot summer day, Tommy Turtle laid on the rocky shores of Knowles, Cat Island. As he laid there, his tongue was dry and forehead dripping sweat. He said to himself, "I am very hot. What can I do to cool my body off?" He thought and he thought, but he didn't know what to do. He decided to ask his friend Woody Pecker.

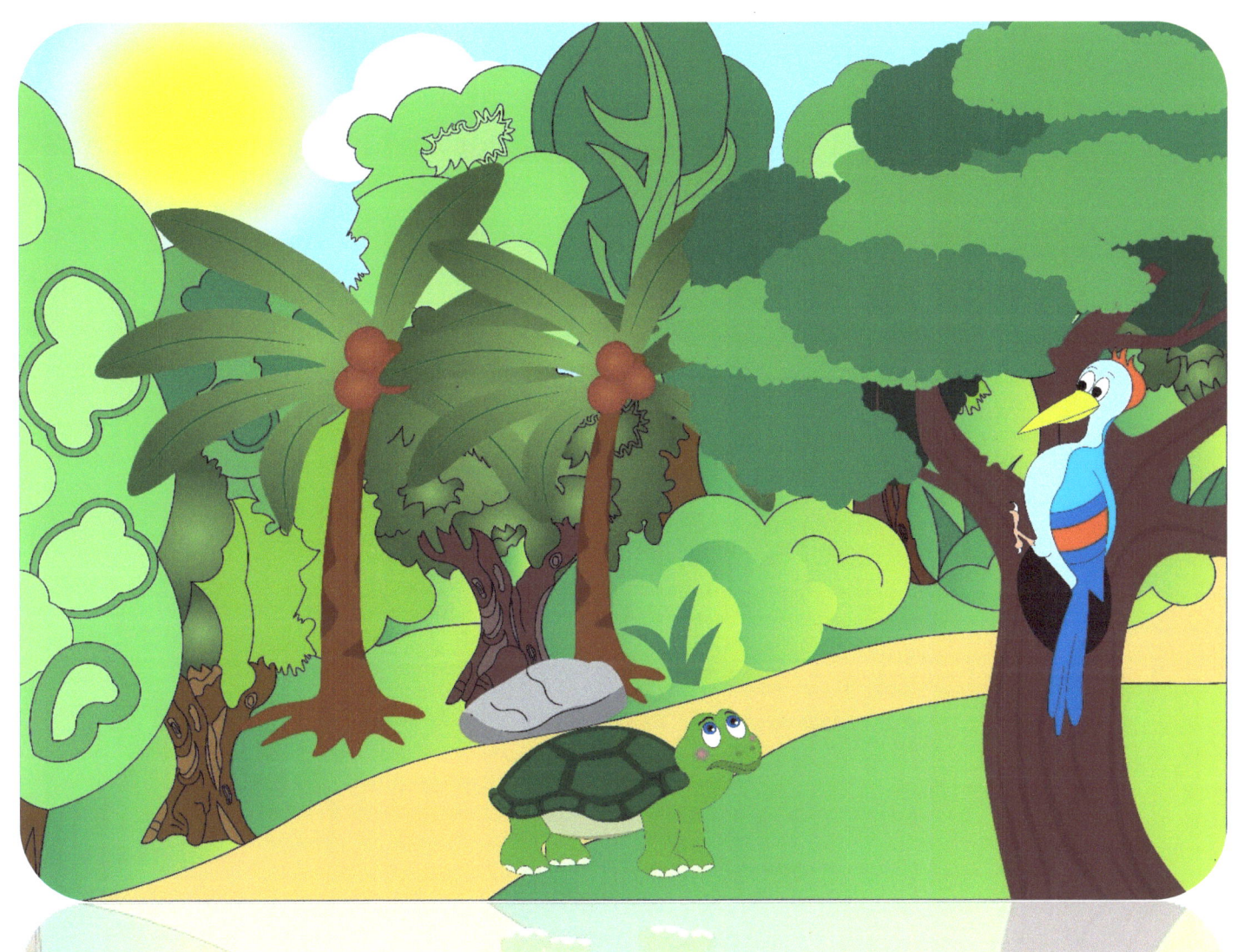

"Good day Woody Pecker," said Tommy Turtle. "I have a problem. Can you help me?"

"I will try," Woody Pecker answered.

"I was sitting on the rocks where the temperature is always good. Now, I am very hot, and I don't know what to do," Tommy Turtle explained.

Woody Pecker said, "I am a bird, and I have feathers. I do not know how to help you."

Feeling sad and still hot, Tommy went down the path.

Tommy went along his way until he met his other friend, Billy the dog.

"I have a problem. Can you help me?" he asked.

"I will try," said Billy.

Tommy began to explain, "I was sitting on the rocks where the temperature is always good to me. Now, I am very hot, and I don't know what to do."

Billy replied, "I cannot help you. I am a dog. I have hair on my body, and I am a warm blooded animal. I am in the class of animal called mammals."

Hearing this, Tommy was disappointed and still very hot.

He went back to sit on the rocks where he met his
friend Freddy Fish. He said to him, "I have a problem. Can
you help me?"

"I will try," said Freddy.

So, Tommy told his story for a third time. "I was sitting on the rocks where the temperature is always good to me. Now, I am very hot, and I don't know what to do."

Swimming slowly near the shore, Freddy Fish said to him,
"I am cold-blooded just like you. When I feel too warm, I
go to the deepest part of the ocean where the water feels
cool to me. However, you have lungs, and you live on land.
So I cannot help you."

Tommy sobbed as he strolled by a pond.

As he stood there, Ken the Frog hopped out the pond and waved to him.

Ken said, "Hi Tommy, why are you crying?"

Tommy sighed and told his story a fourth time. "I was sitting on the rocks where the temperature is always good. Now I am very hot, and I don't know what to do."

I'm so sorry, but I can't help you. I am an amphibian. As an adult, we undergo metamorphosis. We begin our lives in water with gills and tails and as we grow, we develop lungs and legs for life on land. We are cold blooded vertebrate living in water or on land.

Tommy was disappointed once again. He was still feeling very hot and was about to give up.

"Maybe I'm never going to feel cool again," he thought to himself.

Just then, Sally Snake crawled up to him.

"I heard that you have a problem, and I think
I can help you".

"How can you help me?" No one has been able to help me all day," grumbled Tommy.

"Do you know that you and I are reptiles, and we are cold-blooded?" Sally Snake asked.

"Really! What does that mean?"

"Well!" Sally Snake began, "Our temperature depends on the environment. So, it's not very smart to sit on the rocks on a hot summer day."

"Oh, really!"

"Of course not! Staying on those rocks would certainly keep you very hot," Sally replied.

After hearing this, Tommy was feeling embarrassed
that he did not know such important things about
himself.

"Don't worry Tommy. To cool off, you can take a swim
in a pond or just lie down under the shade of a
tree," Sally said to him.

Without wasting another minute, Tommy moved

quickly towards the pond with joy.

He jumped in the pond and swam for a long while.
Once his body had cooled off, he crept lazily onto
the land.

"Ah, now I feel great!" he said with a wide smile.

With these words, Tommy stretched out under a
tall, shady tree and fell fast asleep.

When Tommy woke, he ran quickly to tell his friends that he is not a bird, mammal, fish, or amphibian but a reptile. His body temperature depends on his environment.

Key Words

temperature   stretched

cold-blooded   lazily

warm-blooded   disappointed

lungs   embarrassed

gills   metamorphosis

www.ingramcontent.com/pod-product-compliance
Lightning Source LLC
Chambersburg PA
CBHW041308180526
45172CB00003B/1015